Fluid Power
Educational
Series

Maintenance,
Troubleshooting,
and
Safety
in
Pneumatic Systems

Joji Parambath

Maintenance, Troubleshooting, and Safety in Pneumatic Systems

Copyright © 2022 Joji Parambath

All rights reserved

ISBN: 9798653453434

https://joji.books.com

First Edition: 2020
Revised Edition: 2022

Disclaimer of Liability

The contents of this textbook have been checked for accuracy. Since deviations cannot be avoided entirely, we cannot guarantee full agreement. Only qualified personnel should be allowed to install and work on pneumatic equipment. Qualified persons are those authorized to commission, to ground, and tag circuits, equipment, and systems following established safety practices and standards.

To the loving memory of

Kunnummal Parvathy Teacher

Table of Contents

PREFACE

While a sound knowledge of pneumatics and electro-pneumatics is essential as a foundation, a new understanding of various maintenance practices and procedures is very important for a successful production manager, plant manager, or maintenance technician.

Components used in pneumatic systems can fail in various modes through malfunction, misuse, and ageing. Hence adequate safeguards must be provided to prevent personal injury or damage to equipment in the event of any critical failure.

A single rule that is relevant to all maintenance procedures in all plants and under all circumstances is "be careful". Carelessness and failure to observe safety precautions are two things that maintenance personnel cannot afford.

This book systematically explains all aspects of maintenance, troubleshooting, and safety in pneumatic systems, to make this book useful on the shop floor. A section on energy saving highlights the steps that need to be taken for saving a substantial amount on energy costs.

Enjoy reading the book.
Your feedback is most welcome.

JOJI Parambath

Safety Instructions

To prevent personal injury or damage or both to machinery or equipment during assembly, handling, maintenance, and repair:

- Read carefully and follow the instructions in owners' manuals, service manuals, warning labels, and other instructions.
- Always follow correct procedures and use proper tools and safety equipment.
- Be sure to receive appropriate training, and allow only trained and knowledgeable personnel to install, maintain, or repair machinery or equipment.
- Never work alone while installing, maintaining, or repairing machinery or equipment.
- Always use correct components in applications for which they are approved.
- Set the pressure of compressed air to zero gauge pressure, and stop the flow of compressed air, before starting an inspection
- Do not touch high-temperature areas. Hot surfaces pose the risk of burns if the temperature is > 40°C [100°F].
- Disconnect electrical supply (when necessary) before installation or servicing.
- Disconnect air supply and depressurize all pipelines connected to a component before installation or servicing.
- Be sure to assemble components correctly, and test them for proper assembly and operation before being placed into regular use.
- Operate the machinery/equipment within the manufacturer's specified pressure, and temperature.
- Warnings on the product should not be covered by paint.

[The list is incomplete. More safety tips are given at the end of the book. Please refer to the operator's manual provided by the manufacturer of the machinery or equipment]

Chapter 1 | Maintenance – General Terms and Procedures

In general, the term 'maintenance' covers a broad range of routine maintenance and repair activities, intended to keep a system in a satisfactory working condition. The same broad definition applies to the specialized area of pneumatics as well. Next, we present the classification of maintenance, in general, terms.

Classification of Maintenance

Maintenance can be classified into two basic categories:

- Preventive (or proactive) maintenance
- Corrective (or breakdown or reactive) maintenance

Preventive maintenance is undertaken when a machine is operating correctly to prevent any potential failure of the machine. It is performed regularly, usually as per a schedule or a checklist, for ensuring the efficient working of all components of the machine, at all times. The objective of performing the preventive maintenance of a machine is to prolong its useful service life.

On the other hand, corrective maintenance is undertaken on a machine after its failure. This activity consists of finding the fault and repairing the machine. It is needless to say that preventive maintenance is the most effective maintenance strategy, and one has to focus on the prevention of failure rather than the troubleshooting of the machine.

Therefore, pneumatic maintenance personnel should have the necessary maintenance and troubleshooting skills, apart from the knowledge of the physical laws of pneumatics as well as the functions and symbols of pneumatic components, for performing successful maintenance and troubleshooting activities.

It is also essential to take adequate safeguards to prevent personal injury and damage to pneumatic machines during maintenance and troubleshooting activities.

The following sections explain all aspects of the maintenance and troubleshooting of pneumatic machines and systems.

Definitions of Maintenance Activities

In general, maintenance involves certain closely-related activities, such as inspection, servicing, examination, and overhaul. The following paragraphs clarify the meanings of these maintenance-related activities in respect of a machine.

- **Inspection**: This term refers to the maintenance activity that comprises a careful observation/scrutiny of the machine, usually without dismantling it. This activity typically includes visual and operational checks.

- **Servicing:** This term refers to the cleaning, adjustment, lubrication, and other servicing functions of the machine without dismantling it.

- **Examination:** This term relates to the inspection of the machine with necessary dismantling, measurement, and non-destructive tests to obtain useful information regarding the condition of the components/ subassemblies of the machine.

- **Overhaul:** This term refers to the extensive work done to repair and/or replace the worn-out or defective parts of the machine. The parts are dismantled, partly or wholly. The components, which are worn beyond the acceptable limit, are replaced. The assembly is followed by functional checks and measurements to ensure the satisfactory operation of the machine.

Requirements for Preventive Maintenance

The primary objective of any preventive maintenance activity on a machine is the prevention of its failure or breakdown. Intelligent management of preventive maintenance of a compressed air plant helps increase plant uptime and reduce unplanned shutdowns. The most general requirements to achieve these goals are as follows:

Know the Machine

The first step involved in the preventative maintenance of the machine is to become familiar with the machine. For that, a maintenance technician should know about the layout of the machine, the routing of lines, the functioning of all the components used in the machine, and the operation of the whole circuit. Once the machine is familiarized, regular maintenance becomes easy.

Understand and Follow the Best Maintenance Practices

An understanding of the proper maintenance procedures and the knack to follow them are the pre-requisites for carrying out good preventive maintenance activities. The actual maintenance of the machine must be decided by the complexity of its structure, its operating cycle, and the amount of time available to maintain it.

Compile a Maintenance Checklist

It is also necessary to develop and implement a maintenance checklist or schedule for the machine following the best maintenance practices. A checklist essentially standardizes routine maintenance activities. It lays down the intervals (daily, weekly, monthly, quarterly, and any other period) at which inspection and servicing activities are to be carried out.

Monitor Costs and Performance

The cost-benefit analysis of the preventive maintenance program of a machine should be carried out to see if any corrective steps are needed and to increase the operational efficiency of the machine.

Follow Instruction Manual

Trained maintenance personnel should carefully study the instruction manual of the machine supplied by its manufacturer for compiling its maintenance checklist. The machine manufacturer provides only general guidelines, which cannot be taken as the exact maintenance program conforming to local conditions.

Ensure the Safety

The safety of personnel and machines is paramount while carrying out the preventive maintenance of a machine. Therefore, a maintenance person should follow safe practices during maintenance activities.

Stock Spares

Spares are crucial for maintenance duty and should be available in handy. Hence, it is a good practice to stock essential spare parts of the machine with proper inventory control to facilitate its fault servicing with minimum interruption to production.

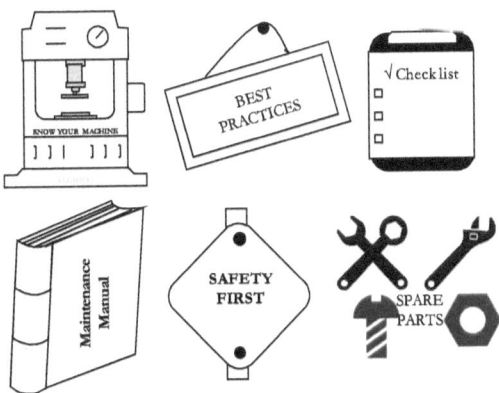

Figure 1.1 | Requirements for preventive maintenance

Figure 1.1 shows the requirements for preventive maintenance schematically.

Air Quality Classification (As per ISO 8573-1: 2001)

ISO 8573-1: 2001 stipulates contaminants and quality classes of compressed air for general use. Air contains solid, water and oil particles as contaminants. A quality class number is defined for each of these contaminants according to the permissible levels of parameters deciding the contamination. These parameters and permissible values of them against each class are given in Table 1.1.

Table 1.1 | Permissible levels of contaminants
as per ISO 8573-1: 2001

Class	Solids		Water	Oil
	Max. particle size (μm)	Max. concentration (mg/m³)	Max. pressure dew point (°C)	Concentration (mg/m³)
1	0.1	0.1	-70	0.01
2	1	1	-40	0.1
3	5	5	-20	1
4	15	8	+3	5
5	40	10	+7	25
6	-	-	+10	-
7	-	-	Not specified	-

A quality class 1.7.1 means that air contains solid particles of a maximum size of 0.1 μm and a maximum concentration of 0.1 mg/m³, an unspecified amount of moisture, and oil particles of maximum concentration of 0.01 mg/m³.

ISO 8573-1 2001 is superseded by ISO 8573-1 2010. The old version is given earlier for easy learning purposes. The air quality classification ISO 8573-1 2010 is given below.

Air Quality Classification (As per ISO 8573-1: 2010)

ISO 8573-1: 2010 stipulates contaminants and quality classes of compressed air for general use. These parameters and permissible values of them against each class are given in Tables 1.2, 1.3, and 1.4.

A. Solid Particles

Table 1.2 | Permissible levels of solid particles (Ref: ISO 8573-1)

Class (A)	Max. particles per m³			Remarks
	0.1- 0.5 μ*	0.5-1 μ*	1-5 μ*	
1	0-20000	0-400	0-10	Highest quality
2	20001-400000	401-6000	11-100	Better than class 3
3	-	6001-90000	101-1000	Better than class 4
4	-	-	1001-10000	Better than class 5
5	-	-	10001-100000	Better than class 6
6	With the mass concentration of $>5\mu$ particles: ≤5 mg/m³			Better than class 7
7	With the mass concentration of $>5\mu$ particles: 5-10 mg/m³			Better than class x
x	With the mass concentration of $>5\mu$ particles: >10 mg/m³			Lowest quality

*Size of solid particles
Note: Class is assigned for the worst-case situation.

An air quality class is specified as a combination of the three air quality numbers in the format: A.B.C.

For example, a quality class 1.2.1 means that in each m³ of compressed air, the particulate count should not exceed 20000 particles in the 0.1-0.5 μ size range, 400 particles in the 0.5-1 μ size range and 10 particles in the 1-5 μ size range.

B. Water

Table 1.3 | Permissible levels of water (Ref: ISO 8573-1)

Class (B)	Vapour pressure dew point	Liquid water g/m³	Remarks
1	≤-70°C	-	Use class 1 desiccant dryer
2	≤-40°C	-	Use class 2 desiccant dryer
3	≤-20°C	-	Use class 3 desiccant dryer
4	≤+3°C	-	Use class 4 refrigerant dryer
5	≤+7°C	-	Use class 5 refrigerant dryer
6	≤10°C	-	Use class 6 refrigerant dryer
7	-	≤0.5	Better than class 8
8	-	0.5-5	Better than class 9
9	-	5-10	Better than class x
x	-	>10	Lowest quality

C. Oil Particles

Table 1.4 | Permissible levels of oil particles (Ref: ISO 8573-1)

Class (C)	mg/m³	Remarks
1	0.01	Highest quality
2	0.1	Better than class 3
3	1	Better than class 4
4	5	Better than class x
x	>10	Lowest quality

Class 0

A quality class 0 can be defined as per a written specification between user and supplier. It should be more stringent than class 1.

Review Questions

1. Define maintenance. How is it classified?
2. What is the importance of preventive maintenance?
3. What are the advantages of preventive maintenance?
4. What are the benefits of employing qualified maintenance personnel for the maintenance of pneumatic systems?
5. What is the most significant cause of pneumatic system failure?
6. What is the importance of instruction manuals supplied by manufacturers for maintenance personnel?
7. What are the factors upon which the intervals of maintenance activities depend?
8. What is the difference between inspection and examination?
9. What is overhauling of equipment?
10. What is the meaning of the quality class 1.2.1 as per the ISO 8573-1 standard?

11. The most general requirements of preventive maintenance of a machine are: (Select the most appropriate choice)
a) Ensuring safety, stocking spares, and taking risks
b) Knowing the machine but carrying out the maintenance when the machine is about to fail
c) Following the best maintenance practices and the maintenance checklist
d) Following manufacturer's instruction manual and troubleshooting only when a fault occurs

12. Air quality classes as per the ISO 8573-1: 2010 standard for solid particles are decided by the permissible levels of: (Select the most appropriate choice)
a) Maximum particle size only
b) Maximum particle count per m^3 for particle sizes up to 5μ only
c) Maximum mass concentration of $>5\mu$ per m^3 only
d) Maximum particle count per m^3 for particle sizes up to 5μ and maximum mass concentration of $>5\mu$ per m^3

Chapter 2 | Preventive Maintenance of Pneumatic Systems

Pneumatic systems are neither perfect nor immune to failures. Faults in pneumatic systems can be attributed to many reasons.

A potential reason, for the faults, is the presence of dust, moisture, heat, etc., which must be removed for the system to function correctly.

Leakage is an essential concern of all compressed air systems, especially the older systems. Leaks are caused by damaged/ corroded pipework, faulty pipe joints, poor quality of fittings, incorrect installation of air-line couplers, faulty seals, dynamic wear, cuts in tubing and open drains taps. Compressed air is a costly medium. Moreover, air leakage is one of the main reasons for pressure drops.

Another reason for faults is the loose bolts and nuts in the system. Each connection should be periodically checked for its tightness and inspection should include checking of possible loose bolts on each component/subassembly of the system. Another critical concern in pneumatic systems is frictional losses.

Faults in pneumatic systems are also due to the stoppage/slower performance (lack of force), poor performance (low speed), erratic operation, and/or leakages.

Therefore, preventive maintenance is the best way to prevent faults for a long time and prolong the life of these systems.

Best Practices for Preventive Maintenance

Concern for Safety
The safety of personnel and equipment should always be kept in mind while carrying out routine maintenance activities. Safety must be built into a pneumatic system by incorporating

interlocks, power-failure locks, and an emergency shutdown feature. Though the responsibility of building safety into a machine rests with the manufacturer of the machine, every engineer/technician should observe and practice relevant safety regulations. Further, every maintenance personnel should make a serious attempt to know the hazards involved in his/her occupation. Proper training should be provided for every maintenance personnel to perform maintenance activities without sacrificing safety.

Important Activities Related to Safety
Some of the safety-related actions that should be followed during maintenance of pneumatic systems are as follows: (1) Wear personal protective equipment, (2) Wear appropriate and safe clothing and footwear, (3) Avoid wearing loose clothing around moving machinery, (4) Shut down the machine before cleaning or repairing, (5) Provide control interlocks or protective guards for clamping devices, (6) Keep safety guards in place, (7) Isolate and shut-off machines from the pneumatic power source before any maintenance or servicing activities, and (8) Ensure that the prime mover has been locked out and tagged so that it cannot be started while carrying out the repair work, (8) Use air fuses to avoid whiplash of long tubing or hoses, and (9) Take adequate safety measures especially near a compressor, for avoiding fire hazards.

Activities for Good Maintenance during the Design Phase
The planning for the good maintenance of a pneumatic system begins right from the design stage of the system. That means, the design must facilitate easy maintenance and the efficient removal of contaminants. Further, it is better to plan and design a pneumatic system with correctly-sized and inherently-safe components for its reliability and long service life.

Activities for Good Maintenance during Installation
Additionally, it is required to install a machine in such a way that there is ample space for the operation and maintenance of the

machine. It is also essential to provide adequate lighting and a clean environment. The service indicators must be easily visible, and service points must be accessible without any difficulty.

Removal of Heat and Solid Contaminants
It is essential to maintain various conditioning devices, like coolers, filters, dryers, and lubricators, at regular intervals, for the removal of heat and other harmful contaminants. The routine maintenance activities generally include cleaning, visual inspection, running checks, and servicing of filters, lubricators, and coolers. The reason for any abnormal noises in the system should be investigated immediately.

Plugging Leaks
An essential requirement of any pneumatic system is to stop leakage, as expensive leaks cost big money. Application of soapy water or commercially available leak detecting liquids, like aerosol sprays, on suspected joints, might reveal the presence of leaks. An ultrasonic leak detection instrument can be used for a quick and precise location of leaks.

Mounting of Components / Machines
The mounting bolts of components must be torqued correctly. Further, every prime mover must be perfectly aligned with the associated load part to reduce undue stress on components, seals, and bearings. A maintenance technician should check for any vibration, loose bolts, or misalignments of components in a system.

Best Practices for Preventive Maintenance – Final Points
When maintaining the same equipment for a while, it is better to be familiar with the equipment. Be observant and careful! Listen, look, and touch a piece of equipment for which one is responsible, for uncovering any developing problems.

The right maintenance procedure can be summed up in a few words: 'Keep the assembly tight, keep the system clean, keep moving parts lubricated, and inspect the system frequently and thoroughly' to perform maintenance and not repair.

Malfunctions in Pneumatic Systems

A general account of malfunctions in pneumatic machines and systems and their causes is given below:

Malfunctions due to Contaminants

A pneumatic system requires compressed air free of moisture and impurities, like dust particles and pipe scales, for its reliable and safe operation, and long service life. Lack of maintenance may fail the system. Various conditioning devices, like coolers, filters, dryers, and lubricators, are installed for the removal of heat and harmful contaminants.

Rust particles are introduced within the piping of a pneumatic system by moisture resulting from condensation. The free-moving particles combined with oil and water sludge can scratch seals and abrade surfaces of precision-made parts of valves and cylinders in the system, thus causing leaks. The particles can also block orifices and cause valve spools to jam. Further, flow passages may become restricted, resulting in a reduced flow rate and increased pressure drop. The moisture present can also wash away lubricants from the valves, resulting in the faulty operation of the valves, corrosive damage to the internal surfaces of actuators and valves, and excessive wear of system components.

Malfunctions due to Under-sized Air Supply

Many a time, pneumatic systems/machines are added to a factory or workshop without enlarging the capacity of the existing compressed air supply or the pipe size becomes too small for the increased flow of air due to the capacity enlargement. As a result, malfunctions can occur sporadically. For example, a sudden pressure drop, caused due to the actuation of nearby components, may result in variations in piston force or speed for a short duration.

Malfunctions due to Under-lubrication/Over-lubrication

The majority of modern pneumatic machines are pre-lubricated. The lubrication of compressed air is required in many older pneumatic systems. It provides lubrication to seals, prevents sticking of moving parts, and controls wear.

The absence of lubrication or under-lubrication in a system will cause increased wear and the consequent deterioration of the system components. Over lubrication may produce the sluggish operation of valves, cylinders, and pneumatic tools.

Malfunctions due to Improper Mountings

A cylinder mounted incorrectly, will produce an undue strain on its mounting plate and mounting bolts.

The piston-rod of a cylinder that is not adequately supported or not correctly aligned with the centreline of the associated load will exert severe strain on its seals and glands. Connections and supports of components that are subjected to vibration should be examined for their tightness.

Consequences of Poor Maintenance

The systematic maintenance of a pneumatic system is necessary to ensure the long service life and reliability of devices and components used in the system.

Lack of regular maintenance may result in the loss of air and associated pressure drops, premature wear of moving parts, and increased downtime of pneumatic components.

These effects eventually result in the increased downtime of the system and shortfalls in production.

Preventive Maintenance of Pneumatic Systems

The preventive maintenance of a system is to be carried out by taking into account the complexity of the individual components of the system.

Review Questions

1. What are the usual causes of the failure of pneumatic equipment?
2. What are the causes of compressed air leakage in pneumatic systems?
3. What are the reasons for friction and vibration in pneumatic components?
4. Explain how safety can be built into a pneumatic system.
5. What early actions can a maintenance technician take to work safely while maintaining pneumatic systems?
6. List five important activities that can be taken during the maintenance of pneumatic systems.
7. What are the activities that can be taken for good maintenance during the design phase of a machine?
8. What are the activities that can be taken for good maintenance during the installation of a machine?
9. Explain briefly the malfunctions caused due to the presence of contaminants.
10. How can a pneumatic system be maintained to reduce solid contaminants?
11. How can a leak in a pneumatic system be detected?
12. How can heat development in pneumatic systems be controlled?
13. Explain briefly the malfunctions caused due to improper mountings.
14. How can a pneumatic system be maintained to reduce vibration?
15. Explain briefly the malfunctions caused due to over-lubrication.
16. Explain briefly the malfunctions caused due to under-lubrication.
17. How can a pneumatic machine be maintained to reduce friction in its internal parts?
18. What are the consequences of poor maintenance in pneumatic systems?

Chapter 3 | Maintenance of Air Compressors

Compressors are the most common energy supply units in pneumatic systems. They are classified into piston type, screw type and vane type compressors. A compressor can be of the single-stage or multi-stage type, reciprocating or rotary type, and oil-injected or oil-free type. A multi-stage compressor is usually provided with an intercooler to remove the heat of compression. It may be noted that the rotor and housing unit of a rotary compressor is also known as airend.

Compressors are invariably coupled to diesel-engine-driven or electric-powered prime-movers directly or through gears or belts. The part that connects the motor to the compressor is usually called a drive train.

A maintenance program should be in place for the care of a compressor. It reduces leakage, limits friction and heat, secures the installation firmly, and avoids unexpected downtime. It can provide years of reliable service, ensure the smooth running of operations, and increase the service life of the compressor. An option to perform the maintenance of a cutting-edge compressor is to engage a professional service specialist. However, with a few simple maintenance activities, as highlighted below, the compressor can run at peak performance.

The first thing to remember while operating, adjusting, or maintaining a compressor is to follow all safety precautions including wearing proper personal protective equipment (PPE). It is essential to disconnect the compressor from the power source and allow it to cool off before commencing any maintenance activity. Also, vent all air pressure in the receiver tank. Next, lock out and tag out the electrical supply and isolation valve on the compressed air piping. The maintenance personnel should be mindful of hot surfaces, rotating parts, hot oil, and high-pressure section of the system. It is also important to receive proper training for carrying out maintenance activities.

The routine maintenance and troubleshooting activities on compressors must be carried out following the manufacturer's instructions. The instruction manual usually contains essential information on the safety guidelines, maintenance instructions, and troubleshooting tips.

A compressor should be located in a clean, accessible area for easy visual inspection, servicing, and examination. Avoid a location where the air is liable to ingress high humidity and absorb a large concentration of solid particles.

Intake filters with suitable micron ratings must be provided at the suction side of the compressor to remove coarse particles from the incoming air. Ensure that the intake filters are clean at all times. If the filters are dirty, then the compressed air remains contaminated and the flow of intake air gets restricted. Therefore, the filter elements must be cleaned or replaced at regular intervals. A filter element can be cleaned by removing dirt and washing it in a suitable flushing solvent. Intervals for cleaning/replacing intake filters are governed by the state of cleanliness of the surrounding air as well as the manufacturer's recommendations. A typical interval can be 500 hours of operation. It is recommended to replace the filter elements with genuine OEM parts. A simplified procedure to inspect and replace air filter elements is highlighted below.

Check/Replace the air filter elements
- Unscrew the filter top from the filter base
- Separate the filter top cover from the base
- Remove the filter element
- Blow out dust and debris from the filter element
- Replace the element, if necessary
- Reconnect filter top to the base

Oil Lubrication
Oil is used in a compressor to maintain lubrication between its internal moving parts. It is also used for cooling the internal parts

of the compressor. Note, the oil should be of the right type having the correct viscosity. Further, it must be maintained in a clean state.

If the viscosity and cleanliness of the oil are not maintained or the oil is not changed at regular intervals, the associated rotors and bearings will be stressed. Therefore, the regular checking of oil in an oil-lubricated compressor is an essential maintenance task. The interval between the oil replacements typically ranges from 4000 to 8000 hours of operation. It is critical to remove all the residual oil in a compressor before filling it with new oil.
The simplified steps for checking and replacing the oil in a piston compressor is given below.

Checking the oil level

- Disconnect the compressor from the power source
- Ensure that the compressor is on a flat surface
- Remove the oil fill cap
- Check the oil level through the sight glass, replenish if necessary
- Inspect the oil in the crankcase assembly
- Check for dirt or debris, if any
- Change oil, if necessary

Replacing the oil

- Drain out the used oil and collect it in a container
- Fill the crankcase with the right type of oil to the required level
- Place back the cap tightly

Check/ Replace Oil Filters

Check oil filters and air-oil separators at regular intervals. Replace oil filters at intervals that can typically range from 4000 to 8000 hours of use. In dirty atmospheric conditions, the filters need to be changed more often.

Cooler / Heat Exchanger, Compressor

A compressor can reach very high-temperature levels in hot, aggressive, and less ventilated environments. The overheating of the compressor can elevate the temperature of discharge air to a high level and cause premature oil failure. The layout of the compressor must support quick dissipation of all the excess heat for its optimum performance. A cooler/heat exchanger can be used for removing excess heat from the compressor.

The inlet and outlet temperatures of the coolant (air/water) in a heat exchanger needs to be regularly checked. Remember, if the heat exchanger is dirty, it will not be able to do its job well. Therefore, it must be visually inspected for signs of contamination and kept clean as often as needed for its optimal performance.

Drive Train, Compressor

As stated earlier, compressors are coupled to their prime-movers directly or through belts. The maintenance of the drive train includes the checking of the direct drive or v-belts.

A direct-drive compressor needs to be aligned perfectly. Any misalignment will cause severe internal stresses and premature failure of airend and motor bearings. Therefore, the alignment of the compressor with the motor should be checked regularly.

The belts in a belt-coupled compressor need to be fitted with proper tension to ensure efficient coupling between the pulleys of connected parts. Over time, the belts tend to wear and crack. A failed coupling cannot transmit power effectively and can cause vibration in compressors. Therefore, it is required to check the belt condition at regular intervals (typically 500 hours), adjust the tension, if necessary, and replace them before they lose their effectiveness.

Roller Bearings, Electric Motor

The bearings on the drive motor of the compressor require periodic lubrication with the right type and amount of grease. The lubrication can prevent the bearings from thermal breakdown and prolong the life of the electric motor. Typically, grease the bearings every 4,000 hours. However, remember that over-lubrication of the bearings is liable to cause premature failure of the part.

Intake Vents

Intake vents in a compressor can get lined with dirt. Therefore, it is necessary to inspect the intake vents weekly and clean them periodically to prevent the dirt from getting sucked into the compressed air.

Piston / Screw / Vane Elements

The reciprocating and rotating elements, such as pistons, screws and vanes, of a compressor are subjected to extreme stress. They can be damaged by friction, overheating, under-lubrication, corrosion, overpressure, and vibration. The deterioration of the rotating elements is indicated by the wearing of rotating elements and bearings, the sticking of vanes, inlet valves, and discharge valves, and leakage due to the weakening of seals. Vanes in a vane type compressor wear out quickly and hence they need to be replaced with new ones every few years.

Check for oil-seal leaks and bearing noise for the symptoms of a potential failure.

Auxiliary Devices

Inspect and adjust compressor controls, safety shutdown systems, condensate drains, filter clogging indicators, and pressure gauges for their correct operation once every month.

Noise Generation, Compressor

The most common causes of compressor noise include worn motor bearings and rubbing of the rotating elements against the housing. Check for abnormal noise and take remedial action to reduce noise.

General Maintenance Activities, Compressor

The level of maintenance required for compressors depends on the type of compressors, auxiliary equipment (such as dryers, filters, and control mechanism used, duty cycle, ambient air conditions, air quality requirements, and choice of safety levels. To summarise, the most common maintenance activities for compressors are listed below:

- Check the oil level
- Inspect for oil leaks, if any
- Fill oil, if necessary
- Replace the oil, if necessary
- Drain condensate in the tank
- Check connections for air leaks
- Check the safety relief valve
- Inspect belts for tightness
- Replace belts, if necessary
- Check air filters, clean or replace
- Check intake valves, clean if necessary
- Check for abnormal noise
- Check for any vibration
- Check and tighten all bolts, however, never over-tight
- Check the safety shut-off mechanism, if any
- Overhaul the compressor

Sample Check List, Compressor System

A sample checklist for the routine maintenance of air compressors is given in Table 3.1:

Table 3.1 | A sample preventive maintenance checklist

Procedure	Daily	Weekly	Monthly	Annually (200 hours)
Overall visual inspection	Y			
Check for oil leaks	Y			
Check the oil level	Y			
Drain condensate in the tank	Y			
Check connections for air leaks	Y			
Check air filters	Y			
Check for abnormal noise	Y			
Check for any vibration	Y			
Check safety guards	Y			
Check the tank for cracks/rust	Y			
Check the safety relief valve		Y		
Clean the cooling surface of the compressor		Y		
Clean the cooling surface of the intercooler		Y		
Apply soapy water around joints during compressor operation and watch for leaks		Y		

Check the lubricant for contamination		Y		
Inspect belts for tightness			Y	
Check intake valves			Y	
Check and tighten all bolts			Y	
Check safety mechanism			Y	
Clean or replace the intake filter				Y
Overhaul the compressor				Y

Y - Yes

Maintenance of Air Receivers

Safety devices on air-receivers like pressure relief valves must be maintained in good functional order. All special pressure vessel rules of air-receivers must be observed fully.

A large amount of water condenses as the air in the receiver tank cools and gets collected at the bottom of the tank. The water can be drained easily using either a manual valve or an automatic drain valve. The drain valve needs to be inspected regularly to ensure that the complete unit is working properly.

Review Questions

1. Why should a maintenance program be in place in a compressor system?
2. What all safety precautions a technician should take while operating, adjusting, or maintaining a compressor?
3. What are the advantages of referring to the instruction manuals of manufacturers while maintaining or troubleshooting a pneumatic system?
4. What are the maintenance activities that can be carried out on the intake filter of a compressor?
5. Explain how a lubricant for a compressor can be maintained.
6. What are the actions that can be taken to monitor heat in a compressor?
7. What are the essential maintenance activities for the drive train of a compressor system?
8. Why is it important to keep roller bearings in good condition?
9. What are the important maintenance issues concerning the moving elements in a compressor?
10. Describe various maintenance activities essential for air receiver tanks.

11. Mark the <u>incorrect</u> statement
(Select the most appropriate choice)
a) The major components of concern for the proper maintenance of compressors are intake filters, lubricants, heat exchangers, and bearings.
b) The reciprocating and rotating elements of a compressor can be damaged by friction, overheating, under-lubrication, corrosion, and vibration.
c) The large water condensate in the receiver tank of a compressor must be drained using either a manual valve or an automatic drain valve.
d) The manufacturer's instruction manual of a compressor usually contains information on the profile, products, and services of the company.

Chapter 4 | Maintenance of Aftercoolers

An aftercooler is a heat exchanger used for removing heat from compressed air discharged from a compressor and cooling the air. It also removes water generated as a result of condensation. It uses either an air-cooled or water-cooled mechanism. An air-cooled aftercooler uses ambient air to cool the hot compressed air. A water-cooled aftercooler is generally a shell and tube type heat exchanger. Compressed air flows through the tubes and water flows through the shell. It also contains a condensate separator.

Compressed air flows through the aftercooler. The cooling media flows over the cooler. As the air cools, the moisture present in it condenses into liquid water. The air with a centrifugal motion then reaches the separator. This motion causes the condensed water and solid contaminants in the air to hit the inner walls of the separator and drip away to the bottom of the separator. The collected condensate can then be drained before it makes any damage to the downstream components.

An aftercooler can be provided as an integral unit in a compressor package. Alternatively, a stand-alone aftercooler is a separate unit installed downstream of the compressor. Most aftercoolers are sized to cool the air to an approach temperature of 2.7°C to 11°C (5 to 20°F) above the ambient air temperature.

An aftercooler should be located directly downstream of the associated compressor. Proper maintenance on the aftercooler will keep its operation most efficient. A dirty aftercooler can result in both warmer air temperatures and increased pressure drop across it.

Installation Points, Aftercooler

Some points to be noted while installing an aftercooler are highlighted below:

- Install the aftercooler close to the outlet of the compressor
- Install shut-off valves on the inlet and outlet sides of the coolant water supply
- Use union joints to connect the coolant water pipes for easy maintenance
- Removable type tube bundle unit may be used for easy maintenance
- Larger after-coolers should be installed with proper support
- Use a drain pipe for the removal of condensate

Maintenance of Aftercoolers

Aftercoolers must be maintained properly for their optimum performance. The essential maintenance activities required for aftercoolers are presented below.

Maintenance, air-cooled Aftercooler

- Inspect and clean the aftercooler regularly
- Check condensate level daily and the condition of the drain valve occasionally

Maintenance, water-cooled Aftercooler

- Inspect and clean the aftercooler regularly
- Check condensate level daily and the condition of the drain valve occasionally
- Check approach temperature and pressure drop
- Monitor the quality of coolant water
- Replace the coolant water, if necessary
- Drain the coolant water under the freezing condition to prevent damage
- Drain the coolant water when it will not be used for a long period
- Clean the inside of the coolant water pipes if necessary

Review Questions

1. Explain the constructional features of an air-cooled aftercooler.
2. Describe the constructional features of a water-cooled aftercooler.
3. What is the general working principle of an aftercooler?
4. What is the significance of the approach temperature for an aftercooler?
5. What are the typical approach temperature values for aftercoolers?
6. Write some installations points concerning aftercoolers.
7. Write two points concerning the maintenance of air-cooled aftercoolers.
8. Write three points concerning the maintenance of water-cooled aftercoolers.

9. Mark the <u>incorrect</u> statement: (Select the most appropriate choice)
a) An aftercooler should be located downstream of the compressor.
b) An aftercooler removes the heat of the compressed air and water generated as a result of condensation.
c) An important maintenance activity concerning an aftercooler is to drain the water condensate regularly
d) Fixed type tube unit may be used in a water-cooled aftercooler for easy maintenance

Chapter 5 | Maintenance of Dryers

Compressed air delivered by a compressor contains a large amount of moisture in the vapour form as well as in the liquid form. The water is capable of damaging the components in a system and causing interruptions and consequent production downtime. Therefore, the moisture should be removed from the compressed air to a level acceptable to the application.

A dryer can be used for removing a large amount of moisture from the compressed air medium of a system. It is a critical element needed to prepare high-quality compressed air. It must provide a stable dew point, maximum reliability, and minimal overall life-cycle costs. Remember, ISO 8573-1 defines the quality classes for compressed air based on the amount or concentration of solid particles, moisture, and oil particles present in the air.

Many types of dryers are designed with different technologies, sophistication, and costs. The most commonly used dryers are desiccant dryers and refrigerant dryers. They are briefly explained below.

Desiccant Dryers

A desiccant dryer uses a tower containing an adsorbent media like silica gel for attracting and adsorbing water vapour. Adsorption is a physical process in which the molecules of water vapour get attracted to and deposited on the surface of the media.

However, the adsorbent media gets saturated with water as it adsorbs more and more water. Fortunately, the saturated silica gel can be regenerated using various methods. Two towers are generally used for non-stop production. One tower (online) can be used for drying the compressed air and the other one (offline) for regeneration at a time. The tower change-over takes place on a time basis or when the silica gel in the online tower, which is drying the air, gets saturated.

Further, the dryer package includes a pre-filter for removing particles of submicron levels, such as oil particles, and an after-filter for removing the dust generated by the drying media. During regeneration in a heated blower type dryer, ambient air is drawn in through an intake filter, passes through a blower, is heated to a high temperature, and passes through the offline desiccant media.

The dryer also uses a set of valves for controlling the air supply at its inlet and outlet, the change-over of the towers for the alternate regeneration and drying, and exhausting the moisture-laden purge air. This set of valves normally includes inlet and outlet switching valves, purge exhaust valves, check valves, and solenoid valves. It also includes drain valves for removing the accumulated condensates.

The adsorbent desiccant material must have optimum adsorption and regeneration capacity to provide a consistent dew point, high crush strength to prevent the breakdown of the desiccant during operation, and low dusting to prevent blockage of downstream filtration. In normal working conditions, the typical lifetime of a good-quality desiccant material is approximately 5 years. Remember to follow a proper procedure while charging the drying chambers with adsorbent desiccant material.

Desiccant dryers are typically designed to provide a constant pressure dew point of -40°C [-40°F] or optionally -70°C [-100°F] to satisfy the requirements of the compressed air quality needed for applications in line with the classes as specified in ISO 8573-1. The achievement of the specified quality class for the application eliminates costly interruption of the production processes due to moisture.

General Precautions, Desiccant Dryers

Remember, dryers are pressure vessels and must conform to the rules concerning pressure vessels in one's region. All safety precautions must be taken during their installation, operation, and maintenance. It may be noted that the owner is responsible for maintaining a dryer in safe operating conditions. The general precautions to be observed before any maintenance, repair work, adjustment or any other non-routine checks on dryers are highlighted below.

- Employ safe working practices and observe all related work safety requirements and regulations.
- Installation, operation, maintenance and repair work must only be performed by authorized or trained personnel.
- Switch off the dryer before carrying out any maintenance on it. Also, disconnect all pressure sources and vent the internal pressure of the system before dismantling any pressurized component.
- The power isolating switch must be opened and locked.
- Parts and accessories shall be replaced if unsuitable for safe operation.

Maintenance of Desiccant Dryers, Introduction

A high-quality dryer has only low maintenance requirements. The dryer must provide easy access to all service-relevant components. Check the dryer for correct operation after any maintenance. Protect the interconnecting tubes/hoses, and maintain them in good condition. It is recommended to entrust the job of replacing the desiccant material in a desiccant dryer to a qualified service technician. Some of the maintenance activities for compressed air dryers are presented below.

Safety Precautions during Maintenance of Desiccant Dryers

- Always use correct personnel protective equipment (PPE) and tools
- Carry out maintenance work on a hot machine only after it is cooled down
- Ventilate the working area adequately
- The maintenance work area should be properly barricaded and a warning sign bearing the label 'Maintenance work in progress' should be displayed
- All regulating and safety devices shall be in operational mode
- Maintain cleanliness during maintenance and repair
- Cover parts and exposed openings with a clean cloth or paper
- Take safety precautions against toxic vapours of cleaning agents
- Make sure that no tools and loose parts are left in or on the machine
- After the maintenance work, set the operating pressures and temperatures.
- Check that all control devices are fitted and they function correctly
- Do not inhale desiccant dust
- Do not overfill a dryer when replacing the desiccant

Tips for General Maintenance of Desiccant Dryers

As stated earlier, dryers are generally maintenance-free. However, certain maintenance activities are to be carried out during the service life of a dryer. Some of the general maintenance activities of a desiccant dryer are listed below:

- Switch off the dryer while carrying out any maintenance activity
- Isolate all pressure sources and vent the internal pressure of the dryer before dismantling
- Use proper technique to tightly charge the drying chambers with adsorbent desiccant material
- Check for the correct operation after the maintenance

Desiccant Filling, Points to Note

The desiccant bed must be filled fully with desiccant materials utilizing all of the available space in the charging bed to achieve its maximum packing density. It may be noted that desiccant attrition can lead to dusting, blocked filters, and loss of dew point. This way of filling prevents air channelling through the desiccant, as found in twin tower designs. Due to channelling, a loosely-filled dryer may require more desiccant to achieve a specified dew point, thus, increasing its physical size and operational and maintenance costs. A dryer that is tightly filled with desiccant materials:

- prevents desiccant attrition
- allows more of the desiccant material used for drying
- allows all of the desiccant regenerated, thus, ensuring a consistent dew point
- provides a low and equal resistance to airflow allowing multiple drying chambers and multiple dryer banks to be used

Disposal of Condensates Generated in Desiccant Dryers

Filters or other discarded elements, such as desiccant materials, associated with a dryer, must be disposed of safely, in line with the local environmental legislation.

The condensate released by a compressed air system and dryers contains tiny particles of oil which are harmful to the environment. Condensate may contain mineral oil aerosols, particles of dust and dirt, cooling and lubricating oil, rust, wear debris, pieces of sealing material, and weld from the pipeline. Therefore, the condensate must be disposed of properly and responsibly. Remember, incorrect disposal of the condensate tends to be detrimental to the environment.

There are many rules concerning the disposal of condensates applicable to the region or country where the dryer is installed, and the rules can vary from region to region or from country to country. Anyone who is responsible for the operation of a compressed air system and dryers must know how to dispose of the condensate properly as per the prevailing rules. Any violation of the rule will incur a fine and cause damage to the reputation of the company.

A floor drain or other facilities should be included in a compressed air plant to handle condensation from the dryers or other equipment. The floor drain must be integrated into the plant in compliance with local legislation.

A Typical Maintenance Schedule, Desiccant Dryers

It is necessary to develop and implement a maintenance checklist or schedule for the dryer following the best maintenance practices. A checklist essentially standardizes routine maintenance activities continuously. It lays down the intervals (daily, weekly, and monthly) at which inspection and servicing activities of the dryer are to be carried out. A typical maintenance schedule for a dryer is given in Table 5.1.

Table 5.1 | A typical maintenance checklist

Period	Activity
Daily	• Check the display panel for status and alerts • Drain the condensate from the pre-filter sump, if an automatic drain facility is not provided. • Monitor drying air temperature
Every 6 months (4000 hours of operation)	• Monitor Dew Point • Check for loose connections • Check for air leaks • Check Switching Valves • Check the pre-filters and after-filters, • Check the air intake filters in the heated blower type dryer • Replace filter cartridges, if necessary • Monitor silencers • Inspect Desiccant
Every five years (40000 hours of operation)	• Monitor Dew Point • Replace desiccant material in desiccant dryers • Check Switching Valve(s), • Replace inlet valves, check valves, etc • Replace elements in pre-filters and after-filters • Replace all O-rings, seals, and nylon washers • Overhaul the desiccant tanks of a desiccant dryer • Replace silencers
Every ten years	• Conduct hydrostatic test according to the relevant standards

Troubleshooting Tips, Desiccant Dryers

The preventive maintenance is carried out on dryers to keep them in perfect working condition at all times. However, faults do occur in dryers, which have to be traced and corrected with minimum delay and expense. The symptoms and possible causes of problems and the remedial actions are given in Table 5.2.

Table 5.2 | A typical Troubleshooting chart for a dryer

Symptom	Possible causes	Remedy
Pressure dew point is high	• The dryer is not regenerated fully	• Close the outlet valve, allow the dryer to regenerate fully
	• Silencers are clogged	• Replace the silencers
	• The condensate in filters is not drained	• Check the drain valve and repair or replace
	• Excessive free water in the dryer	• Check the water separator and drain valve of filters upstream
Dryer leaks from purge valve during the compressor load cycle	Faulty purge valve	Remove and inspect purge valve, clean if required
Build-up	• Air dryer not plumbed correctly • Air leaks	• Connect compressor line to the dryer • Locate leaks, repair

Large noise	Silencers are clogged or out-of-place	• Replace the silencers • Properly fix the silencers
Reduced air delivery	More purge air diverted	Check the associated solenoid valve
Service warning of the timer-controlled LEDs*	Indicating the need for servicing the dryer	Service the dryer

*Typically, orange LED will light up after 7750 hours as a warning and red LED will light up after 8000 hours as an alarm.

Refrigerant dryers

A refrigerant dryer uses the idea of reducing the temperature of the compressed air discharged from the compressor as low as possible to dry the moist compressed air. It typically consists of a heat exchanger, refrigerant compressor, refrigerant condenser, capillary tube, condensate separator, condensate drain, and electrical control unit. The temperature of the air is first reduced in the heat exchanger system and then in a refrigeration unit. The separator unit separates the condensate and the drain part removes it. Refrigerant dryers can dry the air down to a pressure dew point of 3°C.

It may be noted that new dryers should use R-513A refrigerant. This type of refrigerant is a non-ozone-depleting refrigerant and has a very low global warming potential (GWP). Therefore, it causes a significantly reduced environmental impact. This refrigerant is developed to replace the R-134A refrigerant.

Maintenance of Refrigerant Dryers

Weekly
- Check condenser and clean if necessary

Monthly
- Check condenser and clean if necessary
- Drain the condensate
- Open manual drain valve to clean the filter

Every three months or 2000 hours of operation whichever comes first:
- Check drain valves
- Replace heavily contaminated or damaged filter elements
- Check suction pressure

Every six months or 4000 hours of operation whichever comes first:
- Check drain valves
- Clean strainer
- Check suction pressure
- Replace elements in pre-filters and after-filters

Review Questions
1. What are the different types of dryers?
2. What is the basic principle of a refrigerant dryer?
3. What are the main parts of a desiccant dryer?
4. Explain why pre-filters and after-filters are required in desiccant dryers.
5. Explain why many valves are employed in a desiccant dryer.
6. What are the properties required for the desiccant material in a dryer?
7. What is the typical pressure dew point value of the desiccant air drying method?
8. Explain why many valves are employed in a desiccant dryer.

9. What are the properties required for the desiccant material in a dryer?
10. What is the typical pressure dew point value of the desiccant air drying method?
11. What are the general precautions to be taken while maintaining a desiccant dryer?
12. Highlight five important maintenance activities concerning a desiccant dryer.
13. What are the general safety precautions to be taken while maintaining a desiccant dryer?
14. Highlight three important daily maintenance activities concerning a desiccant dryer.
15. What are the important points to be observed while filling the desiccant in a dryer?
16. How can the condensate deposits generated in a dryer be disposed of?
17. What is the typical pressure dew point value of a refrigerant air-drying method?
18. What type of refrigerant is recommended for new refrigerant air dryers?
19. What are the general precautions to be taken while maintaining a refrigerant dryer?
20. Highlight three important maintenance activities concerning a refrigerant dryer.
21. What are the general safety precautions to be taken while maintaining a desiccant dryer?
22. Highlight three important daily maintenance activities concerning a desiccant dryer.

Note:
  (1) A troubleshooting chart for air dryers is given in Chapter 13.
  (2) A case study on the installation points of desiccant dryers are given in Appendix 1

Chapter 6 | Maintenance of Air Mains

Proper maintenance of air mains is essential for the effective removal of contaminants. Condensation can take place in compressed air piping systems. Moisture in compressed air can condense into liquid water when the air is passing through the pipes that are exposed to low ambient temperatures. Therefore, drip legs should be installed at all low points in the piping system.

Another requirement of an air-main network must be to stop leakage, as far as possible. Therefore, regular inspection of the air distribution network for leaks should be taken up, preferably after the close of work. Remember to close all shutoff valves of all air consumers while checking for leaks. Experience shows that 70% of leaks from an air-main network occur in the last few metres of the network, i.e. at or near the take-off point. Any of the methods given below may be used for the detection of air leakage.

1. Application on suspected joints with soapy water or commercially available leak detecting liquids, like aerosol sprays, might reveal the presence of leaks.

2. Next, build up the pressure to the system operating level after closing all consumer lines. Determine the pressure drop at the air receiver over a period (at least overnight). This step indicates the leakage in the system. If more than 10% of the generated compressed air is lost through leakage, it is high time to locate and plug the leaks points in the air-main network.

3. An ultrasonic leak detection instrument can be used to locate leakages. When compressed air leaks, it moves from a high-pressure side to a low-pressure side through the leak site. The expanding air creates a turbulent flow. This turbulence has strong ultrasonic components. The intensity of the

ultrasonic signals falls off quickly from the source, permitting the detection of the exact leak spot.

Review Questions

1. Why maintenance is essential for air mains of compressed air distribution?
2. What is the purpose of fitting drip legs on a compressed air piping system?
3. What are the methods available for detecting air leakage in compressed air distribution systems? Explain any one method.

Note: A troubleshooting chart for pneumatic conductors is given in Chapter 13.

Chapter 7 | Maintenance of Air Service Unit (FRL Unit)

If an air service unit (FRL unit), consisting of a filter, regulator, and lubricator, is not maintained correctly, the investment made on the unit and its installation, turn out to be a mere waste.

A properly sized and maintained filter system can eliminate about 75% of the potential causes of pneumatic power system failure. The following regular maintenance of the FRL unit is of utmost importance:

Filter
- The condensate level in the filter must be checked regularly. It must not exceed the maximum level marked. If the condensate level exceeds the maximum level, the condensate is liable to be drawn into the air stream again. Therefore, the accumulated condensate must be drained, before reaching the maximum level, either manually by opening the drain screw or automatically.

- Ensure that auto drain, if provided, operates properly

- Ensure that air leakage does not occur through the filter

- A filter element should be replaced when it is clogged.

- Gasket/O-ring should be replaced when leakage occurs

Regulator
Usually, regulators do not require regular maintenance, especially during the initial years of their service life. After some years of service, leakage may occur in a regulator due to broken springs or faulty seals.

Lubricator

The oil in the lubricator, if used, is consumed in the process of lubricating the compressed air. Therefore, it is necessary to check the oil level in the lubricator and top up, if necessary.

Review Questions

1. Explain the activities to be carried out for the maintenance of filters?
2. Write one maintenance activity concerning lubricator
3. What is the purpose of an air service unit in a pneumatic system?
4. How is an FRL maintained? Explain.

Chapter 8 | Maintenance of Pneumatic Cylinders

A pneumatic cylinder converts pneumatic power into straight-line mechanical power. Pneumatic cylinders are built in both single-acting and double-acting versions. A double-acting cylinder consists essentially of a barrel, piston, piston-rod, end covers, ports, seals, and cushions.

Cylinders have their maintenance requirements usually given in the manufacturer's manual. As a rule, it is not always necessary to open a cylinder for inspecting its internal condition. Precautions must be taken when installing and servicing pneumatic cylinders that will significantly increase their performance and operating efficiency.

Proper mounting of cylinders onto the machine and coupling of piston-rods are of great importance, as any mismatch will result in stress on the cylinders leading to a reduction in their service life. All mountings must be fastened securely. As soon as force is transmitted to a machine, bearing stresses arise at the cylinder barrel and the piston-rod resulting in high edge pressures on the cylinder bearing bushes and on the piston-rod guide bearings. Increased and uneven stresses may also be developed on the piston seals and piston-rod seals. The different cylinder mounting styles should be studied to determine how they could be installed for the best results.

Causes of Pneumatic Cylinder Failure

Incorrect installation of a pneumatic cylinder can cause undue stress being exerted on it. Improper mounting of the pneumatic cylinder is the most common cause of damage to the wiper ring, cup packing, and bushings. Another critical reason for cylinder failure is the presence of dirt, which might go in between the piston and the barrel. The trapped dirt may score the barrel and seals. As a result, the piston seals might become worn or defective, and the compressed air would leak excessively through the piston.

The air service unit should be fitted immediately upstream of cylinders, as close to them as possible, to keep the wear to a minimum. The use of this unit will ensure that no dirt particles, and moisture harmful to the cylinder, can reach it.

Essential Maintenance Activities for Pneumatic Cylinders
Apart from the general maintenance activities, the following maintenance activities can be carried out on a cylinder to keep it in good working condition:

- Check the piston-rod for straightness. Check for any dents or damages on the rod due to impact forces
- Examine the piston-rod bearing for roundness
- Examine the barrel, the piston and the piston-rod for nicks, scoring, and pitting
- Check the cylinder for worn components
- Check and control the leakages in the cylinder
- Replace piston seals, piston-rod seals and/or piston-rod bearings, if leakages occur
- Align the cylinder and its mating part in line, to avoid side loads on the cylinder
- Check the cylinder mountings periodically for tightness or cracks
- Check for sluggish/erratic operation of the cylinder
- Check for the creeping of the cylinder
- Replace the spring in the single-acting cylinder, if broken

Review Questions

1. What are the important parts of a double-acting pneumatic cylinder?
2. What are the important points to be observed while mounting a pneumatic cylinder?
3. Mention five important maintenance activities to be carried out on pneumatic cylinders.
4. What are the important parts of a double-acting pneumatic cylinder?
5. What are the important points to be observed while mounting a pneumatic cylinder?
6. Mention five important maintenance activities to be carried out on pneumatic cylinders.
7. Investigate the reasons for leakage in a pneumatic cylinder.
8. How can the stress on cylinder seals be reduced? Explain.
9. Investigate the reasons for the sluggish operation of a pneumatic cylinder.

Note: A troubleshooting chart for pneumatic cylinders is given in Chapter 13.

Chapter 9 | Maintenance of Pneumatic Valves

As pneumatic valves are manufactured with tight-fitting delicate parts, contaminants can pose significant problems for the valves. Small amounts of dirt, rust, and sludge can lodge in between the mating surfaces of the valves causing their abrasion, seal damage, and internal leakage. The contaminants are also the reason for the sticking of valves and the plugging of small openings in valves. Therefore, it is essential to have a filtered air supply to ensure that no dirt particles get into these valves. Use piping of non-corrosive types, made of copper or nylon, if possible.

Another probable cause of failures is the jammed springs in valves. It is required to connect a valve to a compressed air supply and operate it to detect this failure. The return movement of the valve must be snappy, and any sluggish movement is indicative of jammed springs. In the case of a valve with an in-built spring that has become weak, remove and replace the spring with a new one.

Pneumatic valves can also be damaged, in particular, their operating mechanisms, by incorrect installation and operation. Excess force on the actuating mechanism, for example, could cause the operating mechanism to become disturbed.

Next, the coils of solenoid valves are affected by vibration, moisture, and corrosion. Such conditions in valves ultimately cause malfunctions in the valves and extensive damage to the valve parts. A coil in a solenoid valve, if damaged, must be replaced with a new one.

While servicing a pneumatic valve, it is essential to know the different parts of the valve and how to disassemble/assemble it correctly. Care should be taken for maintaining the cleanliness of the valve and the surroundings while servicing.

Review Questions

1. What are the reasons for the failure of pneumatic valves?
2. Write two prerequisites for maintaining pneumatic valves.
3. Write a few maintenance activities for pneumatic valves.

Note: A troubleshooting chart for pneumatic valves is given in Chapter 13.

Chapter 10 | Troubleshooting Pneumatics

The preventive maintenance is carried out on pneumatic systems to keep them in perfect working condition at all times. However, faults/breakdowns do occur in pneumatic systems, which have to be traced and corrected with minimum delay and expense.

In general, pneumatic failures can be attributed to the presence of contaminants, the clogging of filters, and loose connections. The symptoms of the failures are generally manifested in the form of the development of excessive pressure drops, heat, and noise. In high-production industries, downtime can result in huge losses.

Unfortunately, fault-finding is often performed in a random manner leading to the replacement of components without proper justification. In many cases, this replacement does more harm than help. Hence, a good troubleshooting strategy needs to be in place for the quick detection and rectification of the faults in the pneumatic systems.

General Troubleshooting Procedure

In general, the fault-finding and repair of a faulty system encompass many activities for minimizing the time and cost involved in the fault-finding and repair. These activities include collecting the information on the fault, analyzing and evaluating the data, localizing the fault, conducting necessary tests and repairing the fault. It is recommended to have a fault-finding strategy, as highlighted in Figure 3.1, to trace and rectify faults in any system. The procedure given may seem to be oversimplified. However, this is the basis upon which a good troubleshooting practice is founded.

The first step in troubleshooting a system that has developed some trouble is to understand its operation and associated circuits. The circuit diagram of the system is virtually indispensable as a troubleshooting aid in all instances of system faults/breakdowns. With complex circuits, time constraints do not permit the trouble-shooter to study the whole circuit. It is beneficial to consult the operator of the system and/or refer to the troubleshooting information furnished by the manufacturer for learning quickly how the system should operate. Make use of every source of information to shorten the time necessary for finding the source of trouble. Once sufficient information is collected and evaluated, visualize all possible root causes of the fault.

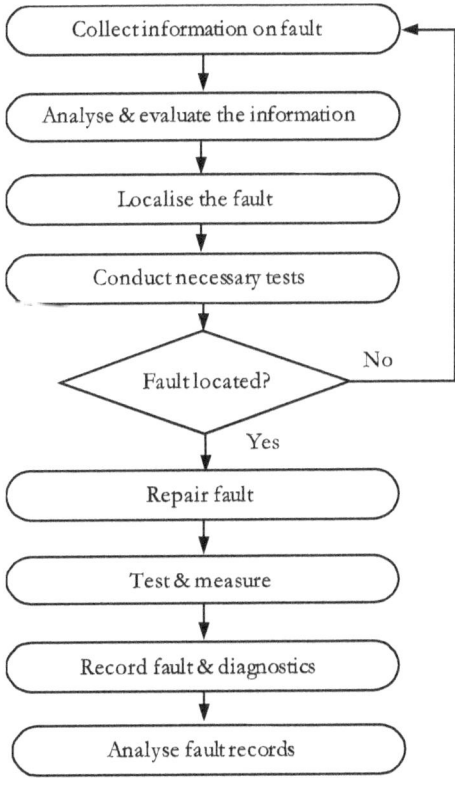

Figure 3.1 | Flow chart of a typical troubleshooting procedure

The most straightforward test may be conducted for finding the section that contains the defective component of the circuit. A careful check of the components involved in this part may lead to the source of trouble. If not, the cycle is repeated until the fault is traced and repaired. The most important rule in troubleshooting a system is to modify only one component at a time. Carry out the circuit analysis with the aid of appropriate test equipment, and not merely by the random checking of components.

The final steps of troubleshooting are concerned with fault recording and fault analysis to discover any recurring pattern of faults or any design and application problem or any shortcomings in the relevant knowledge of maintenance personnel if any. A critical issue to efficient troubleshooting is the proper documentation.

Precautions during Troubleshooting

It is better to follow a step-by-step procedure to troubleshoot a pneumatic system. The precautions to be taken while carrying out troubleshooting activities are listed below:

- The fault-tracing engineer should approach problem-solving in a systematic way. The experience gained by the fault-tracing engineer in every problem-solving exercise is invaluable for solving similar problems in future.

- In a fault-tracing exercise, personal safety and the safety of others is paramount.

- Fault-tracing and repair work should be carried out using approved practices and observing relevant legislation.

- Ideally, all electrical and pneumatic power should be isolated, pressure exhausted, and moving parts mechanically locked.

- Fault finders must keep well clear of the path of all actuators, mechanisms and other hazardous moving parts.

- Electrical equipment should be checked using only the proper test instruments.

Review Questions
1. What is the importance of circuit analysis in troubleshooting?
2. What is the procedure adopted for troubleshooting a pneumatic circuit that has been in existence for a long time?
3. Explain the importance of a good troubleshooting strategy.
4. Explain the standard fault-finding strategy with the aid of a flow chart.
5. Explain the first step of collecting and analysing the information on fault in a pneumatic system.
6. Explain the steps leading to 'localising the fault' in a troubleshooting strategy.

Chapter 11 | Safety in Pneumatic Systems

Nowadays, virtually every industrial production or process equipment employs pneumatic components. How pneumatics is applied becomes very important as it has a bearing on safety. There are only a limited number of standards and regulations for pneumatic systems, and a few of them are listed in the section 'Safety Standards' at the end of this chapter. Therefore, for all factors affecting safety, reference must be made to existing general guidance and regulations from other related engineering fields. Observe all safety procedures when working with pneumatic systems and implement necessary legal requirements for the system. Failure to observe safety regulations could result in injury.

Air receivers used downstream of compressors can store a tremendous amount of energy. The main reason for many of the previous air receiver explosions appeared to have been due to the ignition of oil carried over from the compressor. An explosion is likely to occur if the compressed air gets hot enough to cause spontaneous combustion. When maintaining pneumatic systems, care must be taken in the removal and reconnecting of air-lines. The energy stored in a long pipe or tube will be expelled through its open end in a very short time with enough velocity and force, to cause a severe whiplash of the line. This impact may be avoided by using an air fuse. The danger of stored energy on an inert machine can be extremely dangerous. Where possible, the air should be isolated at two different points, removing the air pressure before disconnection is made.

Pneumatic systems are dangerous but tend to be treated casually. The energy contained in compressed air, the speed and power of pneumatic actuators and unexpected loss of pressure on a clamping device are some of the reasons for this danger. Protection against a loss of pressure on such items as chucks and vices is usually provided for by making them self-locking so that they will be released only if a force is applied in the reverse direction.

One of the most hazardous types of operation is where the operator's hands have to be inserted into the working area to feed or eject a component. The control system for pneumatic clamping devices should be designed and arranged in such a way as to avoid accidental operation. This requirement can be achieved using control interlocks or by using protective guards. If a movable guard is provided, this can be interlocked with the starting of the valve. In case, guarding is not possible, the circuit should be designed in such a way that both hands of the operator should be engaged while the parts are moving. On most control systems safety devices and guards are fitted for the protection of personnel. These should never be manipulated as the safety of personnel may be at risk.

There is a possibility of danger to pneumatic personnel and equipment if the machine stops while under load due to the failure of the air supply. Fail-safe circuits can be designed to protect the operator and/or the machine against electrical power supply failure, air supply failure, overload, and carelessness. Generally, they are designed to return the load to some 'safe' condition should one of these unscheduled events occur. Circuits should be designed, in general, to prevent the inadvertent operation of pneumatic devices. In any fault tracing exercise, the personal safety and safety of others are paramount. Work should be carried out using approved practices and observing the relevant legislation. Any moving parts should be mechanically locked, and trapped air in any section should be exhausted.

In practice, it may be necessary to have a machine or device partially or fully powered up to locate a fault. It may be required to remove and override the guards to gain access to the area of defect. This overriding of guards may present dangers to trouble-shooters, and therefore, they need to exercise great caution and awareness. They must keep well clear of the path of all actuators, mechanisms and other hazardous moving parts. Only service technicians who are qualified to work on pressurized pneumatic systems should install, maintain, and repair them.

Safety Tips
Here are some useful tips for ensuring safety in pneumatic systems

- Train operators and engineers who handle pneumatic systems

- Always wear personnel protective equipment (PPE)

- Disconnect devices from electric supply before doing any troubleshooting work

- Follow the Lockout and Tagout procedures

- Read and follow the manufacturer's instructions for safety

- Use the right diagnostic tools

- Use shutoff valves

- Remove trapped pressure

- Set safety devices on air-receivers like pressure relief valves correctly

- Maintain safety valves in satisfactory functional order

- Mount all components securely

- Do not exceed maximum permissible pressure

- Secure tubing connections before applying compressed air

- Secure hose ends to prevent whipping in case of an accidental cut or break

- Never crimp, couple or uncouple pressurized hose

- Use a tool to actuate a limit switch

- Do not disconnect tubing while under pressure

- Ensure an accessible emergency shutoff valve has been installed in the air supply line

- Never apply compressed air to clothes or skin to clean off dirt and dust

- Ground pneumatic tools correctly to prevent static electricity around flammable vapours and explosive atmosphere

Safety Standards

There are many relevant standards to ensure that a machine is safe. Some of them are listed below:

- ISO 12100 describes the 'Safety of machinery — General principles for design — Risk assessment and risk reduction.'

- ISO 13850 provides the details of 'Safety of machinery — Emergency stop function — Principles for design.'

- ISO 13849-1 stipulates the 'Safety of machinery — Safety-related parts of control systems — Part 1: General principles for design.'

- IEC 60204-1 describes the Safety Of Machinery - Electrical Equipment Of Machines - Part 1: General Requirements

- NFPA 79 provides safeguards for machinery to protect operators, equipment, facilities, and work-in-progress from fire and electrical hazards

Review Questions

1. Write one possible reason for the explosion of compressed air receivers.
2. Explain briefly the stored energy hazard in pneumatic systems. How this danger can be taken care of from the maintenance point of view.
3. What are the safety hazards involved in pneumatic systems?
4. What is the meaning of 'two-hand safety operation' with regards to hazardous types of operation?
5. When are fail-safe circuits employed in pneumatic systems?
6. Write three points concerning stored energy for maintaining safety in pneumatic systems.

7. Safety in pneumatic systems is likely to be compromised by:
a) Wearing PPE
b) Disconnecting the electric power supply
c) Following Lockout/Tagout
d) Ignorance and Carelessness

Chapter 12 | Energy Saving

Compressed air is often considered to be a cheap source of power. It is not. It is comparatively the expensive medium of energy transmission and hence to save running costs and improve efficiency, this medium should not be wasted unnecessarily. Energy is wasted due to many reasons such as air leaks, misuse, excess pressure drops, and over-pressurization. It has been estimated that a typical factory can save up to about 20% of energy by adopting certain easy measures. The following facts may be taken into account for energy saving:

- Use a well-designed pneumatic system. Estimate the needs of air accurately. Design the system for minimum pressure drops and ensure minimum wastage of compressed air.

- As far as possible, employ compressors with an energy management system. These units must be set up in such a way that they can extract cool, dry air, free of fumes and dust particles, from the atmosphere. Care should be taken to install these compressors in cool locations.

- Receivers must be large enough to meet sudden high flow demands.

- It is advisable to condition air through dryers, especially for special applications, like paint spraying and instrumentation. Splitting the main air supply into different sections and conditioning only the quantity required for applications can make excellent savings of energy.

- Energy-saving can be brought about by minimizing pressure drops. This pressure drop can be reduced by using a larger pipe bore, avoiding unnecessary fittings and sharp bends.

- Energy can be saved by using the ring-main installation for the distribution network, preferably with the installed pipes sloping to each corner. This installation can also be provided with a water trap on the dead leg at each corner to collect and remove water. Take-off points must be connected to the top of the main pipe to avoid water pick-up.

- Avoid shutting down the entire system. Use shutoff valves to isolate a section of the installation for servicing and when not in use. The use of shutoff valves is a surprisingly simple and efficient way of preventing energy loss that is often overlooked.

- Leakage is the primary reason for energy loss in compressed air systems. A typical industrial plant may incur a loss of about 25-30% of the compressor output through poorly connected pipe joints, fittings, and couplings. It is often more common to find leakage around fittings and piping. Fixing the air leaks and introducing planned maintenance can produce substantial savings. Introduce a leak prevention scheme by regularly reviewing the system for leaks using ultrasonic leak detection devices.

- Install the FRL unit before each application and pay attention to carrying out its regular maintenance.

- Fine filters fitted upstream of the point of use will reduce cost by protecting equipment against water droplets and solid particles.

- Regularly replacing filter elements can save energy. Use a filter with a service life indicator for the indication of when the filter element must be replaced as a result of clogging. Use a filter with an electrical service life indicator for remote visual and audible warnings of when the filter element is to be replaced. For sensitive applications, a service life indicator can be used to turn off a machine or process automatically.

- A standard 40-μm filter is perfect for a majority of industrial pneumatic applications. Unnecessary use of a 5-μm or sub-μm coalescing element will increase cost due to higher-pressure drop.

- An unnecessary increase in pressure will result in wastage of energy, poor performance, and increased costs due to higher wear and air consumption. Use a pressure regulator to reduce the supply pressure to the optimum level.

- Using lubricated air that will reduce friction and pressure drops can extend the service life of pneumatic tools and equipment.

- Energy-saving begins and ends with measurements. Installation of flow meters and differential pressure gauges can identify the amount of air used and pressure conditions before and after any improvements. With these instruments, the problem areas or deficiencies can be identified.

- Install final control elements as close as possible to the cylinder to reduce the amount of compressed air required for the dead volumes in pipes or tubes leading to the actuator.

- Another primary reason for the wastage of compressed air is the indiscriminate use of air jets as a direct power source, just because it is readily available. Performing air blow operations for cleaning or drying without installing energy-efficient nozzles consumes substantially more energy than necessary. Where compressed air is used directly, selecting correctly sized air nozzles with appropriate control circuits can minimize wastage. Use a regulator to ensure the right pressure of the nozzle. Do not abuse compressed air. Switch off air tools when not in use.

Review Questions

1. What are the ways energy will be wasted in pneumatic systems?
2. How energy can be saved in a pneumatic system by its optimum design?
3. Enumerate energy-saving ideas for compressors and air receiver tanks.
4. Write five steps that can be taken to save pneumatic energy in a typical industrial environment.
5. Explain how energy-saving is brought out by controlling the pressure drops in pneumatic systems?
6. Explain how energy can be saved by the proper layout of air mains.
7. Explain compressed air leakage detection and control for energy saving.
8. How energy can be saved in a pneumatic system by a proper selection of filters?
9. Enumerate energy-saving ideas in pneumatic systems concerning (a) Pressure regulators and (b) Lubricators.
10. What is the role played by flow meters and pressure gauges to save energy in pneumatic systems?
11. How dead volumes of compressed air can be reduced in pneumatic systems?

Chapter 13 | Faults in Pneumatic Systems

Faults in pneumatic systems are generally due to the stoppage/slower performance of a machine (lack of force) or it's poor performance (low speed) or erratic operation or many leakages. The more specific reasons for these faults are:

- Misalignment or mechanical jam
- Power supply failure
- Insufficient pressure or low voltage,
- Twisted tubing
- Burned solenoid coils
- Failure of arc suppression circuits
- Bend piston rods/barrel
- Flow restrictions
- Lack of lubrication
- Insufficient compressed air delivery

Erratic operations can arise from sticking valves or due to any mismatch in the total requirement of compressed air by the system and the actual compressor delivery volume.

If a fault occurs in a pneumatic system, systematic fault tracing is most useful. Repair becomes much more manageable by systematic fault tracing, and above all, the repair time is reduced.

A trouble-shooter should develop the skills necessary to perform a successful troubleshooting procedure.

General Malfunctions

The general malfunctions/disturbances in pneumatic systems, their possible causes and rectification are listed below:

Table 13.1 | Troubleshooting chart for common malfunctions

Fault	Causes	Rectification
Dust, dirt, moisture	Faulty filters, worn seals	-Clean filters -Replace worn seals -Drain and flush fluid
Excess moisture	Faulty drains and traps / Ineffective drying agents	-Drain water traps -Replace drying agents -Check auto drains
Hot air	Faulty cooling system	-Clean cooler -Replace the control valve -Replace cooler
Leakage	Excess clearance in components	-Replace worn seals -Connect pipes and tubing tightly
Excessive wear	Abrasive matter in the air Lack of lubrication High-pressure setting Drive unit misaligned Excessive side loads	-Install correctly-sized filter -Fit air-line lubricator -Lubricate machines -Reduce pressure -Align components properly -Align actuators
Excessive heat	Damaged pump/ motor	-Repair/replace
Excessive noise	Drive motor coupling misaligned	-Align drive assembly -Replace bearings - Replace worn seals
Power supply not working	Damaged compressor	-Repair or replace

Malfunctions in Compressors

Table 13.2 | Troubleshooting chart for compressors

Fault	Causes	Rectification
Excessive heat	-Insufficient compressor cooling -Cooler clogged externally -Low oil level	- Relocate compressor -Clean/replace cooler Fill oil
Low discharge	Low flow	-Set valve correctly -Replace or overhaul -Plug leaking connections -Set drive speed
Compressor slows down or stalls	Low voltage supply	-Provide correct power supply
The safety valve opens too soon	Wrong setting of safety valve	-Adjust the safety valve
Motor turns backwards	Phase reversal	Interchange any two phases
Motor erratic/ not operating	Wrong motor wiring	-Rewire
	Wrong power supply	-Provide correct power supply
	-The drive belt has broken	-Repair
Electrical faults	-Short circuit/ overload	-Rectify fault
	-Switches/ Circuit breaker defective	-Replace
	-Loose connections	-Check the wiring, correct if necessary

Malfunctions in Dryers

Table 13.3 | Troubleshooting chart for dryers

Fault	Causes	Rectification
Excess heating	Insufficient time to regenerate	Regenerate the desiccant completely
Sluggish response	Clogged silencers	-Replace silencers
	Faulty drain	-Check the drain valve
Large noise	Check silencer	-Replace silencer
Poor delivery	Too much purge air	-Check control valve

Malfunctions in Pneumatic Conductors

Table 13.4 | Troubleshooting chart for conductors

Fault	Causes	Rectification
General	Weakened tubing/hose	-Use compatible fluid
	Hose tube cracked	-Protect hose from excessive heat
	Hose bursts	-Use the correctly-sized hoses to withstand shock pressures -Alter environmental or operating conditions -Use correct length hose
	Excessive pressure drop in hose	-Use hose of the correct size -Improve bore condition
	Faulty fittings	-Replace faulty parts
	Leakage	-Replace damaged seal -Correct damaged threads -Use clamps to prevent loosening of fittings

Malfunctions in Pneumatic Cylinders

Table 13.5 | Troubleshooting chart for cylinders

Fault	Causes	Rectification
General	Damaged/worn seals	-Replace seals
	Excessive rod wear	-Prevent wear
	Rod bent/damaged	-Replace rod
	Rod seized	-Repair cylinder
	Faulty alignment	-Align properly
	Broken linkages	-replace linkages
	Loose mountings	-Tighten bolts
	Jerky movement	-Assemble piston packing correctly -Avoid heavy load
	No thrust	-Check pressure -Replace faulty piston -Prevent excessive leakage
	Abnormal thrust	-Bleed cylinder of entrained air -Check and set cylinder cushion
	Excessive jerks	-Align piston-rod and load part correctly
Seal failures	Leakage	-Fit seals properly -Replace worn seals
Erratic operation	Cylinder/valve sticking/binding	-Remove dirt/gummy deposit --Remove worn parts -Install seals properly
	Cylinder internal leakage	- Repair or replace worn parts –Tighten loose packing -Control contamination -Control wear

Malfunctions in Pneumatic Valves

Table 13.6 | Troubleshooting chart for valves

Fault	Causes	Rectification
General	Worn/damaged seals	-Replace seals
	Broken spring	-Replace spring
	Valve leaks	-Replace faulty seals -Replace broken spring
	Broken plunger	-Replace plunger
	Faulty actuation	-Repair actuation system
	Reduced flow rate	-Select correct port size -Clear blocked ports
	Faulty assembly	-Repair or replace
	Short circuit failure	-Check the wiring
Sluggish response	Valve sticky	-Remove dirt -Clean spool
	Blocked vent hole	-Clean vent hole
	Blocked orifice	-Clear blockage -Maintain cleanness
	Degraded internal surface	-Replace defective part
	Restricted exhaust	-Use correct-size pipes and fittings
	Jammed spool	-Check the sealing
	Pressure too low	-Set control pressure
Failure of solenoid coils	Coil hums loudly	-Clean and set right
	Coil vibrates	-Fix coil firmly
	Arc at contacts	-Check arc suppressor
	Coil overheats	-Check coil size
	Burnt out coil	-Replace coil
	The solenoid valve fails to respond	-Check coil -Repair or replace
	Valve chattering noise	-Remove dirt on the seals of valves

Malfunctions in Pneumatic Motors

Table 13.7 | Troubleshooting chart for motors

Fault	Causes	Rectification
Rotates in the wrong direction	Incorrect piping between the control valve and the motor	-Connect circuit correctly
Provides lower speed /torque	Incorrect setting of pressures	-Set pressure correctly
	Compressor delivering insufficient air	-Repair compressor
Fluid leakage	Worn seals	-Replace worn seals

Review Questions

1. What are the three basic types of faults in the pneumatic system?
2. List out a few disturbances or malfunctions in (1) pneumatic valves and (2) pneumatic actuators, along with their possible causes.
3. What are the causes of the following faults in pneumatic systems:
   (a) Compressed air leakage
   (b) Sluggish operation of a valve
   (c) Presence of arc at switching contacts
   (d) Failure of the solenoid coil
   (e) Stoppage of a machine

4. What are the main reasons for wear in components of a pneumatic system?
a)   Abrasive matter in the air
b)   Lack of lubrication
c)   Excessive side loads
d)   All of the above

5. What is the cause of excessive heat in an air compressor?
a)   Low flow
b)   Faulty filter
c)   Loose connection
d)   Insufficient cooling

Appendix 1

Case Study: Installation of Dryers

Precautions during Installation of Dryers
1. Follow the applicable safety regulations
2. Lift the dryer for installation with suitable support equipment
3. Install the dryer where the ambient air is cool and clean
4. Ensure that the air exhausted from the dryer does not recirculate
5. Every distribution pipe/hose must be of the correct size and pressure rating
6. Make sure that all pipes are installed stress-free
7. High-temperature piping is guarded or insulated
8. The dryer must be protected against short circuits by fuses/circuit breakers
9. The dryer must be earthed
10. A lockable power isolating switch must be installed near the equipment.

General Recommendation for Installation, Dryers
1. Install the dryer firmly in level with the floor
2. Provide enough free space [at least 800 mm (2.6 feet)] around the dryer for installation and servicing activities
3. Install a water separator before the dryer to prevent free water from entering the dryer
4. Install an oil/water separator for the draining of pure condensate water, if the condensate contains oil
5. Install a high-efficiency coalescing filter at the inlet of a dryer for removing particles typically down to 0.01 micron and to limit the oil carry-over to a maximum of 0.01 ppm
6. Install a general-purpose pre-filter before the high-efficiency filter for removing particles typically down to 1 micron and to limit the oil carry-over to a maximum of 0.5 ppm

7   Install a dust filter at the outlet of the dryer for removing dust particles originating from the desiccant typically down to 1 micron

8   Install bypass pipes with shut-off valves over the filters to isolate the filters during service operations without disturbing the compressed air delivery

9   Install a pressure relief valve on each tower of the dryer to relieve excess pressure when ball valves are installed at the inlet and outlet of the dryer

Initial start-up of a Dryer

The following points may be taken into account for the initial start-up of a dryer for the first time or after a long period of standstill (more than 3 months).

1. Wear a dust mask, safety glasses, and ear protection
2. Open the bypass valves, if any, of the dryer
3. Shut off the air supply to the dryer
4. Close the outlet valve, if any, of the dryer
5. Remove silencers to prevent them from getting clogged by the desiccant
6. Allow the air supply to the dryer slowly
7. Check connections of the dryer for air leaks and plug leaks if found
8. Switch on the dryer
9. Operate the dryer for several hours with the outlet valve closed
10. Refit the silencers, if removed
11. Slowly open the outlet valve

Normal start of Dryers

The following procedure may be followed for the normal startup of a dryer.

1. Close the inlet valve to shut off the air supply to the dryer
2. Close the outlet valve of the dryer
3. Slowly open the inlet valve
4. Switch on the dryer
5. Gradually open the outlet valve of the dryer
6. Close all bypass valves, if any, of the dryer

Operational Phase of Dryers

1. Check the status of LEDs on the control panel of a dryer at regular intervals

Stopping Dryers

The following procedure may be followed stop a dryer.

1. Open the bypass valves of the dryer
2. Close the inlet valve of the dryer
3. Close the outlet valve of the dryer
4. Operate the dryer idle for a period for depressurizing the vessels
5. Switch off the dryer

15 | References

1. Article on 'AIR COMPRESSOR MAINTENANCE: AN IN-DEPTH GUIDE' by Tim Seberger, Industrial Air Compressor Maintenance: What You Need To Know (rasmech.com)

2. Article on 'Air Compressor Maintenance' Air Compressor Maintenance Guide (portlandcompressor.com)

3. Article on 'AIR COMPRESSOR MAINTENANCE' https://kaishanusa.com/blog/air-compressor-maintenance/

4. Article on 'Industrial Air Compressor Preventative Maintenance' Quincy Compressor, https://www.quincycompressor.com/industrial-air-compressor-preventative-maintenance/

5. Article on 'Maintaining Your Rotary Screw Compressor' by enelson@compressor-pump.com, https://www.compressor-pump.com/author/enelsoncompressor-pump-com/

6. Article on 'Marine Air Compressor Maintenance – Things You Must Know About' by Shalabh Agarwal, Marine Air Compressor Maintenance - Things You Must Know About (marineinsight.com)

7. Article on 'OIL-LESS ROTARY VANE VACUUM PUMP AND COMPRESSOR OPERATION AND MAINTENANCE MANUAL' Gast Manufacturing Inc. 2300 Hwy. M-139 Benton Harbor, MI 49023

8. Article on 'Reciprocating Air Compressor Maintenance' by the Compressed Air & Gas Institute, Reciprocating Air Compressor Maintenance | Compressed Air Best Practices

9. Article on 'Rotary screw air compressor maintenance' by Spencer Hall, Sullair, Rotary screw air compressor maintenance (plantservices.com)

10. Article on 'Rotary Screw Compressor Maintenance Guide' Quincy Compressor, <u>Rotary Screw Compressor Maintenance Guide | Quincy Compressor</u>

11. Article on 'Rotary vane compressor maintenance made easy' by Steve Downes, Ingersoll Rand

12. Article on 'Safe pneumatic system design' by Mary Gannon, Pneumatic Tips, A Fluid Power World Resource

13. Article on 'Safety: Pneumatic Systems' Essentra plc 2020

14. Article on 'Screw compressors: Operation and Maintenance' by TMI Staff & Contributors' <u>Screw compressors: Operation and Maintenance (turbomachinerymag.com)</u>

15. Article on <u>'The Importance of Preventative Maintenance for Air Compressors</u>' The Titus Company

16. Article on 'THE MOST COMMON COMPRESSED AIR DRYING METHODS' VMAC Global Technology Inc., Canada, <u>https://www.vmacair.com/blog/common-compressed-air-drying-methods/</u>

17. Article on 'WATER-COOLED AFTERCOOLER INSTALLATION AND MAINTENANCE INSTRUCTION MANUAL' C A G Technologies, Ontario, Canada

18. Document on 'BENEFITS OF PROPER AIR COMPRESSOR MAINTENANCE' Compressed Air Systems, 9303 Stannum Street Tampa, FL 33619-2658 USA, <u>Home - Compressed Air Systems, Inc.</u>

19. Document on 'Compressed Air Engineering – Basic principles, tips and suggestions' KAESER Compressors, <u>www.kaeser.com</u>

20. Document on 'D Series Single Tower Deliquescent Dryers' VAN AIR SYSTEMS, www.vanairsystems.com
21. Document on 'Desiccant Air Dryers: Heatless, Heated and Heated Blower' Ingersoll Rand
22. Document on 'Oilless Rotary Vane Compressor Operating and Maintenance Instructions' PENTAIR, 2395 Apopka Blvd., Apopka, FL.
23. Document on 'PDNSG-1 Pneumatic Division Safety Guide', ISSUED: 1st August 2006, Supersedes: 1st June 2006, Pneumatic Division Richland, Michigan 49083
24. Instruction Book on 'Heatless adsorption compressed air dryers', Atlas Copco, www.atlascopco.com
25. Joji P., Pneumatic controls, Wiley India Pvt Ltd, New Delhi, 2008
26. Ralph-Christoph Weber, Erwin Orendi, 'Safety in Pneumatic Systems – Workbook TP 250', Festo Didactic GmbH & Co. KG, 73770 Denkendorf, 2011
27. Joji Parambath., 'Compressed Air Dryers' Kindle Direct Publishing

Fluid Power Educational Series Books

1. Pneumatic Systems and Circuits -Basic Level (In the SI Units)
2. Industrial Pneumatics -Basic Level (In the English Units)
3. Pneumatic Systems and Circuits -Advanced Level
4. Electro-Pneumatics and Automation
5. Design of Pneumatic Systems (In the SI Units)
6. Design Concepts in Pneumatic Systems (In the English Units)
7. Maintenance, Troubleshooting, and Safety in Pneumatic Systems
8. Industrial Hydraulic Systems and Circuits -Basic Level (In the SI Units)
9. Industrial Hydraulics -Basic Level (In the English Units)
10. Hydraulic Fluids
11. Hydraulic Filters: Construction, Installation Locations, and Specifications
12. Hydraulic Power Packs (In the SI Units)
13. Power Packs in Hydraulic Systems (In the English Units)
14. Hydraulic Cylinders (In the SI Units)
15. Hydraulic Linear Actuators (In the English Units)
16. Hydraulic Motors (In the SI Units)
17. Hydraulic Rotary Actuators (In the English Units)
18. Hydraulic Accumulators and Circuits (In the SI Units)
19. Accumulators in Hydraulic Systems (In the English Units)
20. Hydraulic Pipes, Tubes, and Hoses (In the SI Units)
21. Pipes, Tubes, and Hoses in Hydraulic Systems (In the English Units)
22. Design of Industrial Hydraulic Systems (In the SI Units)
23. Design Concepts in Industrial Hydraulic Systems (In the English Units)
24. Maintenance, Troubleshooting, and Safety in Hydraulic Systems
25. Hydrostatic Transmissions (HSTs) (In the SI Units)
26. Concepts of Hydrostatic Transmissions (In the English Units)

27. Load Sensing Hydraulic Systems (In the SI Units)
28. Concepts of Load Sensing Hydraulic Systems (In the English Units)
29. Electro-hydraulic Proportional Valves
30. Electro-hydraulic Servo Valves
31. Cartridge Valves
32. Electro-hydraulic Systems and Relay Circuits
33. Practical Book: Pneumatics - Basic Level
34. Practical Book: Electro-pneumatics - Basic Level
35. Practical Book: Industrial Hydraulics – Basic Level
36. Programmable Logic Controllers and Programming Concepts
37. Compressed Air Dryers

For more details, please visit: **https://jojibooks.com**

About the Author

Joji Parambath has been a trainer in Pneumatics, Hydraulics, and PLC for over 25 years. During his career, he has trained numerous professionals from the industries as well as faculty members and students of engineering institutions.

At present, he is the key trainer at Fluidsys Training Centre, Bangalore, India, (https://fluidsys.org), which provides training in Pneumatics and Hydraulics. He has already written two books on Pneumatics and Hydraulics. The publication of the present series of 32 books is intended to restructure and update the existing books.

The author wishes to thank all trainees for their lively interaction and many useful suggestions during the training programmes that prompted the author to write the present series of books. You may send your feedback to joji.p@hotmail.com

10th June 2020